Sponsored Technology Workshops 2019

Held at the 2019 AIChE Annual Meeting

Orlando, Florida, USA
10 - 15 November 2019

ISBN: 978-1-7138-0550-2

Printed from e-media with permission by:

Curran Associates, Inc.
57 Morehouse Lane
Red Hook, NY 12571

Some format issues inherent in the e-media version may also appear in this print version.

Printed with permission by Curran Associates, Inc. (2020)

AIChE
120 Wall Street, FL 23
New York, NY 10005-4020

Phone: (800) 242-4363
Fax: (203) 775-5177

www.aiche.org

Additional copies of this publication are available from:

Curran Associates, Inc.
57 Morehouse Lane
Red Hook, NY 12571 USA
Phone: 845-758-0400
Fax: 845-758-2633
Email: curran@proceedings.com
Web: www.proceedings.com

TABLE OF CONTENTS

586280 Transforming the Engineering Classroom with Digital Data Toolsets

The undergraduate student classroom is rapidly evolving, with new education methods being introduced within the variety of departments and fields taught at major US Universities. One of the biggest changes seen in this field is the shift to digital content and education methods, whether utilizing Blackboard to store and distribute classroom content or introducing digital textbooks for student use. In the engineering classroom this has included the introduction of process engineering toolkits, 3-dimensional modeling software, and increasingly complex calculation methods for the solution of higher order mathematical systems. In order to empower the engineering workforce to address the challenges they are solving, Elsevier has introduced several products to enable the understanding of complex engineering principles or content. 'Engineering Village' is a powerful search and discovery platform that provides data analytics tools for the analysis of engineering literature, including a numeric search tool which allows for searches of not just concepts but specific values of interest as well. 'Knovel' is a technical reference solution that provides answers and insights sourced from digitized and searchable engineering reference content, including critical reference titles and property databases. Together these tools are used across the Chemicals Industry to speed up research, obtain access to needed engineering answers, and promote innovation through exposure to related concepts thanks to the Engineering Thesaurus which ties the concepts together. In this session we will introduce these tools and discuss how they can be used in the engineering classroom to better prepare for the transition to graduate school or the engineering workforce.

585132 Maximize Uptime with Operations Analysis and Prescriptive Maintenance

There is a lot of talk these days about IIoT, Industrie 4.0, Digitalization, with Big Data, analytics, and machine learning. Not initiatives, these are technologies that support initiatives. In this session we'll explore how Operational Excellence is the real initiative, learn the supporting performance indicators, and explore maximizing uptime as the cornerstone of any attempt to improve operational excellence.

We'll take an in-depth look at examples of how prescriptive maintenance can provide early advance warning of failures, and guidance to identify the root causes of damage. With this knowledge, operations and maintenance staff can adjust conditions to eliminate the causes of most failures – acting with confidence to avoid process-related damage and consequent unplanned production downtimes. Or when a shutdown is unavoidable, at least plan the shortest, safest service and repair to mitigate the consequences.

585000 Electrolyte Modeling Basics Simulation

The Electrolyte Modeling Basics workshop is designed for attendees interested in learning about electrolyte chemistry concepts and the fundamentals of modeling electrolyte solutions. A lecture and hands-on OLI Studio and Flowsheet software exercises will be used to demonstrate how complex electrolyte systems can be modeled. Attendees will leave with a qualitative picture of how ions and molecules behave in water and a better understanding of solution properties such as alkalinity, pH, activity coefficients, and more. This course is appropriate for those new to OLI technology, professionals working with this software and those looking for a refresher course in its underlying theory. This course will provide a comprehensive overview of how electrolyte modeling can enhance process simulation results. Case studies such as refinery overhead corrosion simulation, RO membrane simulation for water treatment, and solvent extraction for metals and mining, among others, will be discussed.

586281 Utilizing Search Tools to Speed up the Literature Review Process and Stay Ahead of the Curve

Graduate research requires a broad knowledge of an underlying research field's literature, which can in many cases span decades of work. The effort going into these literature reviews can be significant at the front end of a graduate school career and can often make moving into new research areas or testing new applications for current research seem daunting for even experienced researchers. In order to help ease the understanding of the underlying research literature, as well as see how these areas connect to other research goals, Elsevier has a suite of tools which can be used to help make the literature more manageable and the underlying trends more apparent. In this session, we will discuss how to use these tools to promote overall understanding of the existing literature and find spaces where gaps exist, as well as improving knowledge of the underlying science required for valuable contribution in the lab. Specific topics will include utilizing faceted search strategies to begin a broad literature overview, numeric filtering for very detailed application searches, and generating search strategies to achieve the right results quickly using the tools 'Knovel' and 'Engineering Village'.

584998 XPure- How to Achieve Purity, Productivity and Profitability Demands of Process Industry Using Simulated Moving Bed Technology

Complex streams in process industries require efficient downstream technologies. Simulated moving bed (SMB) chromatography has been known as a potential alternative to process multi-component streams. In the current workshop, we will focus on understanding the state of the art XPure SMB technology. This technology will then be assessed based on key industrial drivers like flexibility, operability and scalability. Finally, some tested and proven case studies will be presented. These cases will describe the impact of XPure SMB technology on purity, productivity and profitability requirements of specific industrial processes.

585127 Aspen Knowledge Accelerates Business Innovation

Innovation occurs when knowledge inspires the adoption of new technologies and work processes that drive improvements in yields and net production. In this session learn how our new offering, AspenTech Knowledge, provides the learning best practices that transform employee competence. Explore how AspenTech Knowledge accelerates adoption of new tools and technologies to improve operational excellence. Get ready for the future trends; see how implementing the AspenTech training certification program measures proficiency and assesses an engineer's potential for success. AspenTech offers many training programs that develop employee proficiency – customizable to unique company needs. Also, see how other companies use existing state grants to fund employee education.

586282 Capital Asset Lifecycle Management (CALM) for Chemical Industry

CALM addresses the challenges of greenfield and brownfield capital projects, such as rampant delays and cost overruns, often caused by lack of system interopérability, low labor productivity and a failure to embrace digitalization. These challenges negatively impact the operational efficiencies of plants and facilities for decades.

According to McKinsey & Company, 98% of megaprojects incur cost overruns or delays, averaging a cost increase of 80% and delaying delivery by 20 months. In addition lack of accurate information results in operational inefficiencies through the life of the plant which is anywhere from 30-50 years. These numbers are staggering and provide a ripe opportunity for the application of Siemens solutions. In this session, you'll learn about the industry's major trends and challenges while conveying the power of digitalization and demonstrating its value throughout the lifecycle of plant or facility. You'll gain an understanding of Teamcenter for Capital Asset Lifecycle Management and how, through the dissemination of engineering and asset data, it can be used to spur your customer's digital transformation.

584999 XPure- How to Design, Build and Implement Integrated Technologies like Expanded Bed Adsorption in SMB Mode

Bio-based industries need selective and efficient downstream solutions to process complex streams like fermentation broth. As fermentation streams contain both dissolved and suspended impurities along with large amount of water (> 70 wt%), it is required separate the impurities and also concentrate the product. In the current workshop, we will discuss the role of EBA technology as a smart separation step for bio-based streams and its impact on product purity, productivity and yield. We will then discuss the benefits of performing EBA in SMB mode and the approach to design, build and implement EBA-SMB. The design will focus on understanding the sensitivity of EBA and SMB technologies to different process parameters. The outcome of sensitivity analysis will then be used as design input for building a suitable and scalable hardware, software package for EBA-SMB integrated technology. Finally, a tested and proven case will be presented to describe the approach of implementing EBA-SMB and the resulting impact on process efficiency and economics.

586283 Simulation for Performance Twin for Process Industry - Process Systems Enterprise Sponsored Technology Workshop

Understanding of transport processes (fluid flow, heat transfer & mass transfer) is key for design, troubleshooting and exploring optimum operating conditions. Computational fluid dynamics (CFD) and particle modeling with Discrete Element Method (DEM) have been identified as key enabling technologies in finding solutions to many of the challenges that surround scale-up; capable of reducing operating costs across manufacturing and quality divisions. We will then use this basis to show examples of how this can be used to solve problems as well as demonstrate the use of Simcenter STAR-CCM+, a multi-physics simulation platform from Siemens. In this seminar we will look at successful deployment of such methods with industrial case studies.

We will also talk about how CFD & DEM simulation can be integrated in to process models/process simulation tools for a deeper insight. This portion will be presented by a guest speaker from Process System Enterprise (PSE)

All presenters will be available to answer questions.

586551 The PSE Academic Teaching Highway (PATH) Initiative – using gPROMS FormulatedProducts for Digital Design of Pharmaceutical Manufacturing Processes

Register Now for this Free Workshop: https://www.psenterprise.com/events/americas/2019/pse-at-aiche-annual-meeting-2019

The PATH initiative aims to align the teaching of process modelling and related digital design techniques in universities with the needs of industrial employers, through the development and provision of software-agnostic teaching materials suitable for incorporation into undergraduate courses. This workshop will give a global overview on this initiative followed by a selection of a hands-on session on digital design of hydrogenation, crystallization and/or dry granulation using gPROMS FormulatedProducts. The workshop will provide an overview of how mechanistic modelling can be deployed to a number of pharmaceutical applications and will include the following elements:

- Introudction to the PATH initiative
- Introduction to the unit operation modules within gPROMS FormulatedProducts
- Simulation of hydrogenation, crystallization and/or dry granulation processes **(Hands-on)**
  - Flowsheeting environment
  - Manual scenario analysis of the impact of Critical Process Parameters (CPPs) on Critical Quality Attributes (CQAs)
- Introduction to Global System Analysis for conducting uncertainty & sensitivity analyses
  - Assessing the impact of CPPs on CQAs for each unit operation**(Hands-on)**

586180 Digital design of robust formulated products and their manufacturing processes through mechanistic modelling

Register Now for this Free Workshop: https://www.psenterprise.com/events/americas/2019/pse-at-aiche-annual-meeting-2019

Join PSE and industry experts for a hands-on software demonstration of gPROMS FormulatedProducts, an innovative and integrated platform for the digital design of robust formulated products and their manufacturing processes. The software enables a consistent methodology and workflow to be applied across the formulated products industries, using mechanistic models for process understanding, design space exploration, tech transfer, optimization, risk assessment, and product performance evaluation.

Scientists and engineers in the pharmaceutical, food, consumer goods, and specialty and agro-chemical industries will choose hands-on modules that demonstrate model configuration, validation with experimental data, optimization, and sensitivity analysis for a particular application. Modules include synthesis and fluid separation, crystallization, wet milling, filtration, continuous direct compression, wet and dry granulation, spray drying, oral absorption, and end-to-end flowsheet modelling.

AUTHOR INDEX